Terrific TOYS

And How They Are Made

By Noah Leatherland

BookLife PUBLISHING

©2023
BookLife Publishing Ltd.
King's Lynn, Norfolk
PE30 4LS, UK

All rights reserved.
Printed in China.

A catalogue record for this book is available from the British Library.

PB ISBN: 978-1-80505-124-4

Written by:
Noah Leatherland

Edited by:
Rebecca Phillips-Bartlett

Designed by:
Amelia Harris

FSC MIX Paper from responsible sources FSC® C113515

All facts, statistics, web addresses and URLs in this book were verified as valid and accurate at time of writing. No responsibility for any changes to external websites or references can be accepted by either the author or publisher.

Image Credits

All images are courtesy of Shutterstock.com, unless otherwise specified. With thanks to Getty Images, Thinkstock Photo and iStockphoto. Cover – Ganna Demchenko, Nadiia Korol, Will Thomass, Anna Kaminska. Used on all pages – Shmelkova Nataliya. 2 – Kolpakova Daria, Monkey Business Images. 4 – StockImageFactory.com. 5 – Emese, Mcimage. 6 – David Leshem. 7 – Zyabich. 8 – lovelyday12, Carlos andre Santos. 9 – Proxima Studio. 10 – pogonici, Watercolor_Art_Photo. 11 – Florida Chuck, mdbildes. 12&13 – Mauro Rodrigues. 13 – Prokrida. 14 – Sergey Ryzhov. 15 – Alexandra Morosanu. 16 – FabrikaSimf, Monkey Business Images. 17 – Frame Stock Footage. 18 – Hodoimg, asharkyu. 19 – Larina Marina, Kyrylo Glivin. 20 – Kolpakova Daria. 21 – Jeanette Virginia Goh, KaliAntye. 22 – StepanPopov, Photo Oz. 23 – Ilina Yuliia, Brookhaven National Laboratory (BNL) [Public domain].

Contents

Page 4 — Terrific Toys
Page 5 — Making Toys
Page 6 — Made by Hand
Page 8 — Wonderful Wood
Page 10 — Soft and Stuffed
Page 12 — Metal Makers
Page 14 — Plastic Pieces
Page 16 — Vivid Video Games
Page 18 — Printing Toys
Page 20 — Make Your Own!
Page 22 — Fun Facts
Page 24 — Glossary and Index

Words that look like this can be found in the glossary on page 24.

Terrific Toys

Toys are things that you use to play. Toys bring lots of fun, but where do they come from? These toys did not just appear. They were made somewhere by someone.

What is your favourite toy?

Making Toys

There are lots of different types of toys and games.

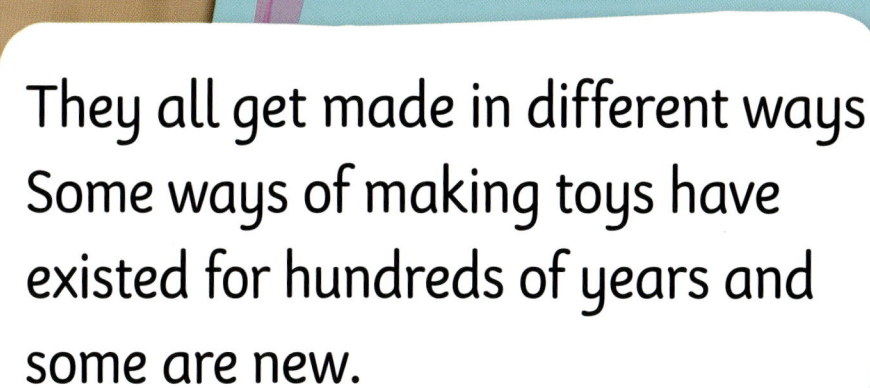

They all get made in different ways. Some ways of making toys have existed for hundreds of years and some are new.

Made by Hand

Toys have been made for thousands of years. Today, most of our toys are made with machines.

Before these machines were **invented**, humans had to make toys by hand.

This toy is thousands of years old!

The oldest toys were made from anything people could find, such as rocks, wood and plants. Even though people have invented lots of clever machines, some toys are still made by hand.

Wonderful Wood

Wood comes from trees. First, a tree needs to be cut down.

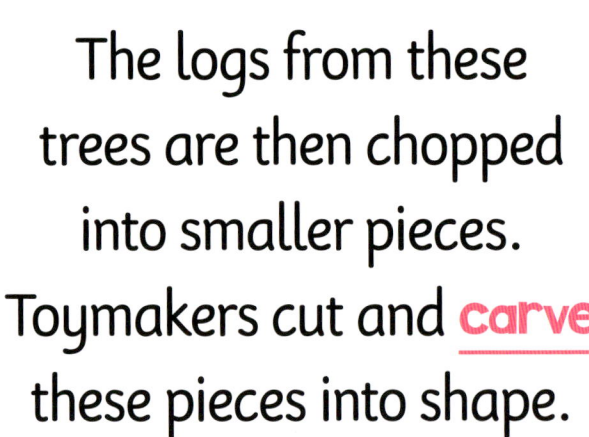

The logs from these trees are then chopped into smaller pieces. Toymakers cut and **carve** these pieces into shape.

Once the wooden pieces have been carved into shape, they are painted and put together. Toymakers might need to use glue or the pieces might fit together just right!

Do you have any wooden toys?

Soft and Stuffed

From teddy bears to dolls, soft toys can look like anything. First, a *pattern* needs to be drawn. The shapes of that pattern are drawn onto big sheets of *fabric* and cut out.

Lots of different fabrics can be used to make soft toys.

10

The fabric pieces get **sewn** together to make the shape of the toy.

The toy gets filled with fluffy stuffing to make it soft. Now, the soft toy is ready to play!

Stuffing

11

Metal Makers

Metal toys are made in a few ways. Some metal toys are made by cutting out parts on a thin metal sheet. These parts are bent into shape and put together to make a toy.

Bending metal into shape is the oldest way to make metal toys. Toymakers have found lots of new ways to make metal toys.

Metal can also be melted and made into new shapes to make toys.

Plastic Pieces

Plastic toys are made using a *mould*. First, small pieces of plastic are melted down into a *paste*. Then, the melted plastic is poured into the mould.

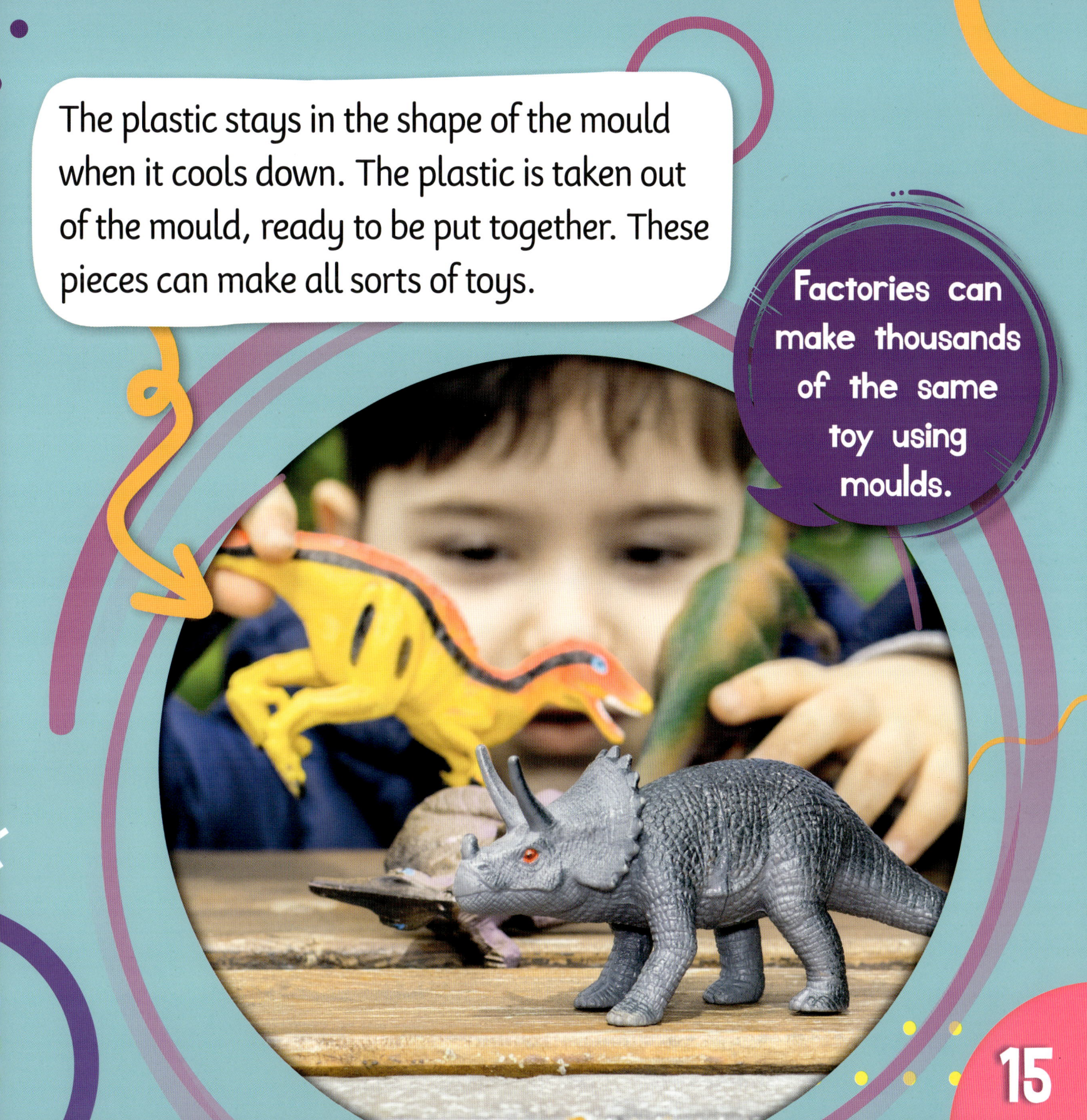

The plastic stays in the shape of the mould when it cools down. The plastic is taken out of the mould, ready to be put together. These pieces can make all sorts of toys.

Factories can make thousands of the same toy using moulds.

Vivid Video Games

Lots of video games are played on a machine called a console. Inside of these consoles are lots of computer parts. The games themselves are **digital** and are made on computers.

Video games can be made by one person or hundreds of people.

There are lots of steps to making a video game. Using computers, video game **designers** plan how their games will look, sound and play. It can take years to make one game.

Do you have a favourite video game?

Printing Toys

A new way to make toys is using a 3D printer.

A 3D printer is a machine that can make three-dimensional shapes, such as cubes and spheres. These are shapes that can be held in your hands.

As well as simple shapes, 3D printers can print all kinds of things.

Some people use them to print different plastic parts that can be made into toys.

19

Make Your Own!

There are lots of different ways to make toys and lots of different **materials** to make toys from. You do not need machines or tools to make your own toys!

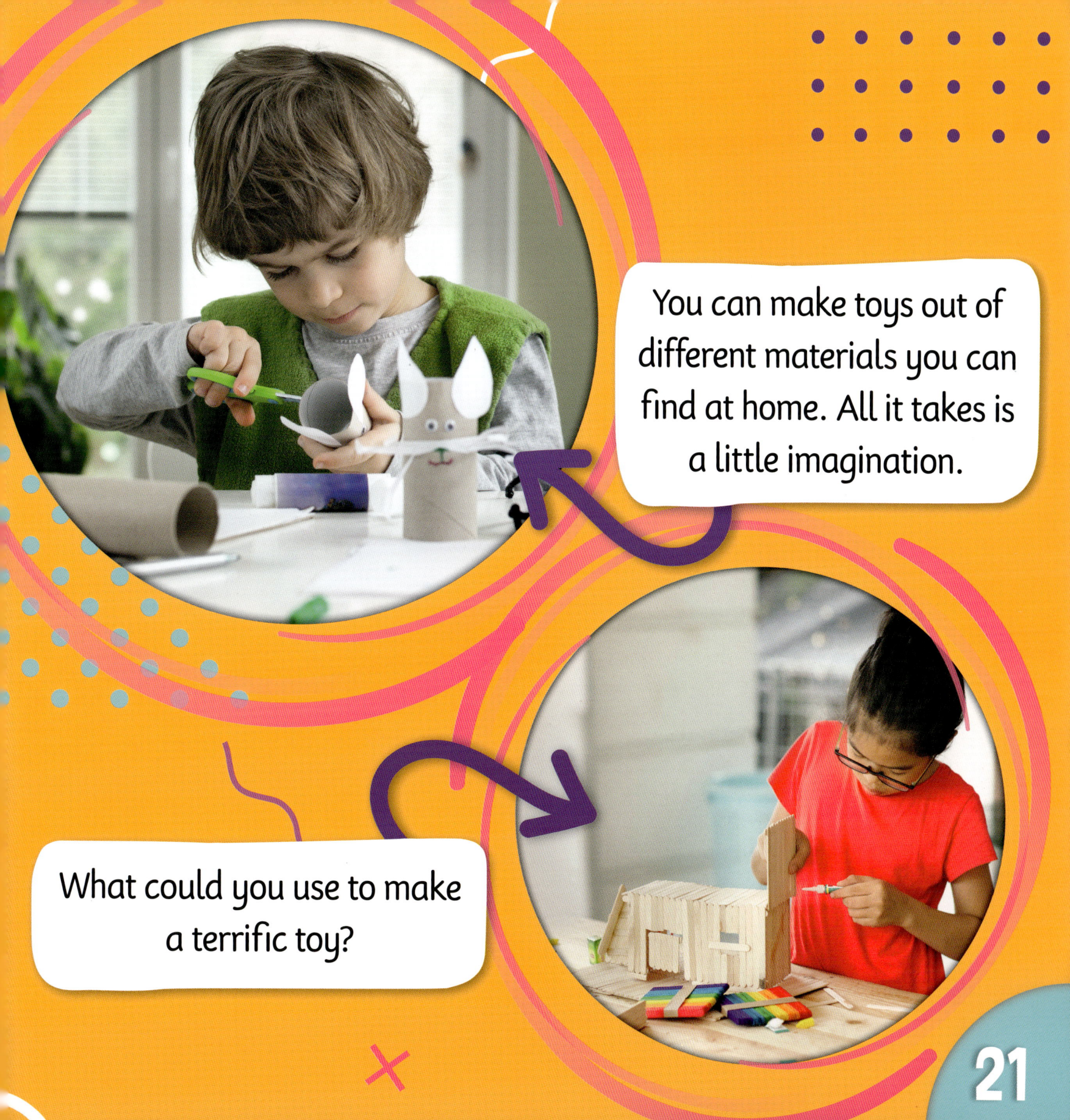

You can make toys out of different materials you can find at home. All it takes is a little imagination.

What could you use to make a terrific toy?

Fun Facts

LEGO makes more tyres than actual car makers.

Some of the earliest toys in history were made of stone.

22

Some toys get made by accident! Playdough was made to clean walls.

The first video game was made in the 1950s.

23

Glossary

carve	cut a hard material into shape
designers	people who plan how something should be made
digital	showing information as an electronic image
fabric	materials made by weaving threads of cotton, wool, nylon, silk or other threads
invented	when something new is made
materials	things from which objects are made
mould	a container with a certain shape that liquids are poured into
paste	a thick, soft substance
pattern	a design used as a guide
sewn	joined pieces together using thread and a needle

Index

computers 16–17
machines 6–7, 16, 18, 20
metal 12–13
moulds 14–15
plastic 14–15, 19
stone 22
video games 16–17, 23
wood 7–9